Smart Starters
Math

Motivational Exercises
to Stimulate the Brain

by Imogene Forte & Marjorie Frank

Incentive Publications
Nashville, Tennessee

Illustrated by Marta Drayton
Cover by Geoffrey Brittingham
Edited by Patience Camplair

ISBN 0-86530-642-7

1   2   3   4   5   6   7   8   9   10   08   07   06   05

PRINTED IN THE UNITED STATES OF AMERICA
www.incentivepublications.com

Table of Contents

Introduction

What is a Smart Starter?

A Smart Starter changes extra moments in a class setting into teachable moments. They are designed take short amounts of time. However, Smart Starters are NOT short on substance. The Smart Starters in this book are packed full of important skills to practice and polish or to reinforce or extend.

When are Smart Starters used?

As their name suggests, they are good for igniting learning. Instead of the slow move into the class period, lesson, or school day, a Smart Starter quick-starts the action. Each one warms up the brain with a sparkling challenge. Students need this kind of spark at times other than the beginning of the day or class period. Use a Smart Starter any time there is a lull, or any time students need a break from a longer activity. They work great to stimulate thinking at the beginning, end, or middle of a class period, or any other time you can squeeze in an extra ten minutes.

Why use Smart Starters?

They're energizing! They're stimulating! They're fun! They nudge students to focus on a specific goal. They "wake-up" tired minds. They require students to make use of previously acquired knowledge and skills. Because of their short length, they give quick success and quick rewards—thus inspiring confidence and satisfaction for learners.

How to Use This Book...

Kick-Off a New Unit

The starters are grouped by math strands. One or more of them might help to ease students into a new area of study. For instance, start off a unit on fractions with *Pizza Problems (page 41)*, or a geometry unit with *Geo-Hunt (page 56)*. Or, use *24 Hours at a Glance (page 70)* to introduce students to the usefulness of statistical graphing.

Spark a Longer Lesson

Any one of these short activities can be expanded. A starter can inspire your students to develop more questions along the same lines—expanding the warm-up into a full-blown math lesson.

Review a Concept

Dust off those rusty skills with a Smart Starter. For instance: Have students been away from the study of measurement for a while? Refresh what they know about units of measure with *Unit Wisdom (page 62)*. Or, strengthen their knowledge of writing mathematical expressions with *How Do You Say It? (page 76)*. Any of these Smart Starters will help to reinforce concepts previously introduced.

Charge-Up Thinking Skills & Ignite Creativity

The Smart Starters are not only for math class. Use them any time to stimulate minds. Doing a Smart Starter will sharpen thinking processes and challenge brains. In addition, Smart Starters work well as starting points for students to create other (similar) questions and problems.

Do You Speak Math?

Math has a language all its own. To do math well, it is important to know the terms. Divide students into small groups. Give each group this list of math terms. Ask them to write a brief definition of as many terms as they can in ten minutes. At another time, share the definitions. Research any terms which the class could not define.

1. perimeter
2. area
3. ratio
4. equivalent
5. greatest common factor
6. quadrilateral
7. perpendicular
8. variable
9. frequency
10. rectangle

11. pyramid
12. logic
13. face
14. composite number
15. congruent
16. vertex
17. rhombus
18. median
19. axis
20. terminating decimal

21. integer
22. square root
23. pi
24. transversal
25. symmetry
26. integer
27. tangent
28. isosceles
29. statistics
30. coordinates

Beware Big Numbers

Large numbers can be confusing. This exercise combines sharp number-reading skills with sharp listening and sharp thinking. Write the following numbers in large numerals on long pieces of paper. Hang them around the room, labeled A-J as shown.

A. 44,404   B. 444,014   C. 404,414   D. 400,004,004   E. 44,440

F. 44,040,040   G. 441,004   H. 4,040,014   I. 40,444   J. 400,040,044

Then, read the spelled-out number listed below to the students. Read each one twice. Give the students twenty seconds to match the two forms of each number. They should identify the number by its label.

1. four hundred four thousand, four hundred, fourteen

2. four hundred million, forty thousand, forty-four

3. forty-four thousand, four hundred four

4. four million, forty thousand, fourteen

5. four hundred forty-four thousand, fourteen

6. four hundred million, four thousand, four

7. four hundred forty-one thousand, four

8. forty-four million, forty thousand, forty

9. forty thousand, four hundred forty-four

10. forty-four thousand, four hundred forty

Hats in Order

Assemble a batch of simple hats for practice with examining and ordering numbers. Prior to class, cut 50 strips of heavy paper, each 25 inches long. Fasten the strips with staples or paper clips, and they become hats.

Use a large, bold pen to write the following numbers on the hats. Keep the hats grouped according to the groups shown below.

1)	0.25;	5,795;	5.9;	5.09;	61,001;	60,101;	5,579;	509;	–3;	2.05
2)	99,099;	9.99;	9.09;	–99;	909,009;	99.99;	–9.9;	909;	99.9;	–0.9
3)	1,234;	12.3;	12 ½;	0.4;	1.2034;	1,324;	–4321;	⅔;	14.32;	1.03
4)	8,888;	8,188;	8,989;	989;	8,888.8;	88.8;	89.08;	998;	99.08;	8,988
5)	–16;	–23;	36;	3.65;	–51.05;	831;	0.016;	–½;	813;	–83

Pass out each group of hats to a group of ten students. Give them one minute to get their hats in order from least to greatest.

Prime Hunt

Group students into pairs. Give them five minutes to search for "primes." Challenge them to find at least fifteen prime numbers in the classroom.

They must keep a list and description of these numbers. They may count items, ask classmates' ages, look at the clock or calendar, and so forth. However, no more than two numbers can come from the same source. (For example, they may not just take all the numbers off the calendar.)

After five minutes, ask students to do the same hunt for composite numbers.

Examples:

PRIMES	COMPOSITES
today's date — 11th	school address — 378
teacher's age — 43	length of hallway — 112 feet
number of boys in this class — 13	weight of math book — 24 oz

Place Value Lineup

Pin the digits 0, 0, 0, 1, 3, 5, 5, 6, 8, 9 on ten different students. Ask them to stand in a line in front of the class. Call out one of the questions below. Give students thirty seconds to line up so that they correctly display the numeral or to conclude that the numeral cannot be formed.

1. Can you show: *one million, five hundred fifty-nine?*

2. Can you show: *five hundred eight thousand, three hundred nineteen?*

3. Can you show: *ten million, nine hundred thousand, five hundred thirty-nine*

4. Can you show: *nine hundred fifty-five million, eight hundred thirteen thousand, six hundred?*

5. Can you show: *fifty thousand, three?*

6. Can you show: *three hundred eighty-six thousand, ninety?*

7. Can you show: *one hundred thousand, sixty-five?*

8. Can you show: *five million, five hundred one thousand, nine hundred sixty-eight?*

Note: Sometimes not all digits will be needed. This can be done with fewer digits and fewer students, or with different digits.

Mysterious Numbers

Follow the clues to find a number that could be the mystery number. Write the clues on cards or on the board to share with students. Students can work in groups to find the mystery numbers. (There may be more than one correct answer.) Let students make up other number mysteries.

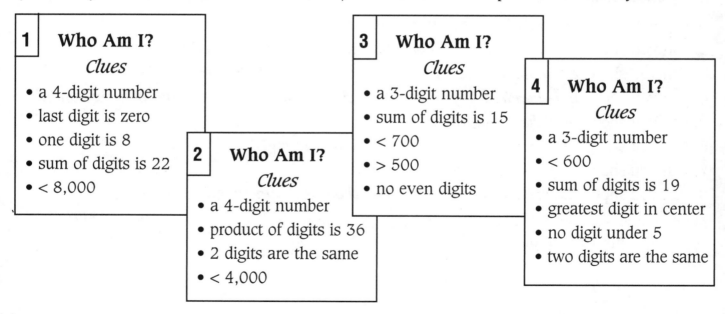

1 Who Am I?

Clues

- a 4-digit number
- last digit is zero
- one digit is 8
- sum of digits is 22
- < 8,000

2 Who Am I?

Clues

- a 4-digit number
- product of digits is 36
- 2 digits are the same
- < 4,000

3 Who Am I?

Clues

- a 3-digit number
- sum of digits is 15
- < 700
- > 500
- no even digits

4 Who Am I?

Clues

- a 3-digit number
- < 600
- sum of digits is 19
- greatest digit in center
- no digit under 5
- two digits are the same

Quick Change, Rearrange

Use this fast-paced challenge to show the flexibility of digits and numbers.

Tell students to record these three digits: 6, 0, 9. Then, read the following instructions. Allow about fifteen seconds for students to write a number for each instruction. The numbers may only use the digits named. Some answers may use only two of the three digits. A digit can be used more than once.

Tell students to write a number . . .

- that is less than 100
- that is a multiple of 5 (use all 3 digits)
- the product of whose digits is 0
- is a multiple of 15 (use all 3 digits)
- is an even number
- is between 6,609 and 6,909

- is a multiple of 96
- that is greater than 600
- the product of whose digits is 54
- is greater than 900
- is a multiple of 6 (use all 3 digits)
- is an odd number

Start over with three other digits. Let students create the instructions to form new numbers.

Speedy Roundup

You will need a stopwatch for this task. Write each number on the board. See how quickly students can round each number to the nearest: 10; 100; 1,000; 10,000; and 100,000.

Numbers:

A. 123,456	F. 483,108
B. 887,654	G. 101,101
C. 585,858	H. 666,666
D. 260,972	I. 709,264
E. 797,979	J. 347,922

Example:

463,726

nearest 10	463,730
nearest 100	463,700
nearest 1,000	464,000
nearest 10,000	460,000
nearest 100,000	500,000

Toothpick Numerals

Pass out toothpicks to help students sharpen their skills at writing Roman numerals.

Group students into pairs. Give each pair fifteen toothpicks. Read each of the following instructions. Give students thirty to forty-five seconds to form the Roman numeral. Check that each pair has formed the numerals correctly.

Show:

A. today's date

B. your age

C. year of your country's independence

D. the number 2,694

E. year of one student's birth

F. number of students in the class

G. answer to: MMMCMXCV — MMDCC

H. the year 1649

Let students take turns creating new problems to solve.

Extraterrestrial Notations

Scientific notation is a shorthand way of writing very large numbers. Quickly review the way scientific notation works. Ask students to rewrite the following "outer space" calculations in scientific notation.

1. Approximate distance from the Sun to Mercury: ....36 million miles

2. Approximate distance from the Earth to Mars: ....49 million miles

3. Saturn's approximate diameter: ....75 thousand miles

4. Approximate (average) distance from Saturn to the Sun: ....88.7 million miles

5. Approximate distance from the Pluto to Uranus: ....1.9 billion miles

6. Neptune's approximate diameter: ....31 thousand miles

7. Approximate temperature of Sun at its center: ....30 million degrees F

8. Number of light years to cross the Milky Way: ....130 thousand

9. Approximate number of stars in the Milky Way: ....200 billion

10. Approximate number of asteroids in the asteroid belt: ....100 thousand

Exponent Examination

Review the concept of exponents with students. Then, show them these equations. Students must decide whether or not each equation is correct.

Answer the question for each: *is it equal or not?*

A. 4^4 = 256

B. 10^6 = 1,000,000

C. 3^5 = 243

D. 33^2 = 66

E. 160^5 = 1,600,000

F. 4^9 = 6,561

G. 2^8 = 256

H. 7^7 = 49,000

I. 6^5 = 7,776

J. 12^4 = 20,736

K. 100^3 = 100,000

L. 40^4 = 2,560,000

To Catch a Thief

Turn students into detectives. Ask them to figure out how much money each thief has in his or her bag and who has the most money.

1. Sly Sam's bag has: 6 bills (None are the same.), 600 quarters, and 1,200 nickels.

2. Slick Sally's bag has: 20 of each kind of bill and 1,035 quarters.

3. Sneaky Sid's bag has: 114 $100 bills, 59 $20 bills, 500 $50 bills, and 150 $5 bills.

4. Crafty Kate's bag has: 450 quarters, 1,515 $50 bills, and 300 $20 bills.

5. Greedy Greg's bag has: 1,260 bills (equal numbers of bills—all under $20) and 2,500 dimes.

6. Pilfering Paul's bag has: 11,000 quarters and 900 $20 bills

7. Selfish Sid's bag has: 10,000 nickels, 30 $5 bills, and 250 quarters.

8. Unlucky Lucy's bag has: 21 of each kind of bill.

How Much Peanut Butter?

You will need a large jar of peanut butter and plenty of crackers.

Note how many ounces of peanut butter are in the full jar. Then, spread a cracker with peanut butter for each student. Eat the crackers, and note how much peanut butter is gone from the jar. Ask each student (or group of students) to estimate how many crackers can be "buttered" from this jar. (Don't forget to count the ones already eaten.)

After the estimates are recorded, get busy spreading peanut butter until the jar is empty. Check the guesses against the actual number of crackers spread. If there are too many for the class to eat, share them with another class.

Smart Guesses

An estimate is not just any guess. It is a careful, or educated, guess.

Discuss this concept with students. Talk about the kinds of information they might gather in order to make careful (not wild) estimates. Then, divide students into small groups and give each one an estimation task. The group should make an estimate, record the estimate, and complete the counting or calculation to find the accurate amount. Compare actual amounts with the estimates.

Estimate ....

...how long would it take for your entire class to get drinks at the water fountain.

...how many shoelaces it would take to make a string long enough to stretch all around the classroom.

...how many television programs the students in your grade watched last weekend.

...how far a student in this group walks (inside the school) to get to all of his or her classes in one day.

...how many pages (total) are in each student's major textbooks.

...how many years (total) all the teachers in the school have been teaching.

How Many Factors?

Start off with a quick review of factoring. Make sure students are clear on the process of identifying the factors in a number. Then, make five charts or sections on the board. Label them with large numbers: 2 Factors, 4 Factors, 6 Factors, 8 Factors, or More Factors. Find the factors for each number below. Count the factors. Write the number on the chart that corresponds to the number of its factors. (For example, write the number 12 on the chart labeled 6 Factors.) Students can work alone or in groups. *Hint: Always remember to count 1 as a factor.*

12	28	50	11
18	99	1000	27
20	7	49	72
35	60	66	121

> **For example:**
>
> 88 has 8 factors:
>
> 1, 2, 4, 8, 11, 22, 44, 88

Make Sense of Multiples

Each of the following pairs of numbers demonstrates multiplication by a multiple of ten. Have students identify which multiple was used to change the first number into the second. The answer will be one of these: ten, one hundred, one thousand, ten thousand, one hundred thousand, or one million.

A. 4,300 ⟶ 4,300,000

B. 16.246 ⟶ 162,460

C. 12.9 ⟶ 12,900,000

D. 88,000 ⟶ 8,800,000

E. 90.909 ⟶ 909.09

F. 450 ⟶ 45,000

G. 83,300 ⟶ 833,000

H. 3.005 ⟶ 300.5

I. 16.101 ⟶ 161,010

J. 909,009 ⟶ 9,090,090

K. 0.3066 ⟶ 30,660

L. 77.077 ⟶ 770.77

Divisibility Quiz

There are many divisibility shortcuts. Describe the two below to students. Then, show them the lists of numbers. Ask them to work quickly and answer *YES* or *NO* to each of the questions on the charts, using the divisibility shortcuts. (Duplicate the charts or write the numbers on the board.)

> A number is **divisible by 8** if the last 3 digits are divisible by 8.
>
> A number is **divisible by 9** if the sum of its digits is divisible by 9.

Divisible by 8?	*Divisible by 9?*
A. 6,473,682	G. 7,265,376
B. 90,901	H. 396,040
C. 1,293,666	I. 8,807,368
D. 55,512	J. 496,277
E. 8,807,368	K. 101,101
F. 12,992	L. 123,321

Which Property?

Review the properties of numbers and operations *(commutative property, associative property, identity property, distributive property)*. Write each of these items on the chalkboard. Ask students to identify the property that is shown by the equation.

A. $3{,}600 \times 1 = 3{,}600$

B. $10 + (48 + 17) = (10 + 48) + 17$

C. $82{,}364 \times 0 = 0$

D. $(44 \times 12) \times 60 = 44 \times (12 \times 60)$

E. $20 \times (30 + 5) = (20 \times 30) + (20 \times 5)$

F. $1{,}500 + 33 + 89 = 33 + 1{,}500 + 89$

G. $29{,}467 + 0 = 29{,}467$

H. $75 \times 3 \times 38 = 38 \times 3 \times 75$

Palindrome Trick

A palindrome is a number, word, phrase, or sentence that reads the same forwards and backwards. (For example: 4,994 or level.) Enjoy this wonderful math trick, which always leads to a palindromic answer. Here's the way it works.

Choose any number:	635
Reverse it:	+ 536
Add:	1,271
Reverse the sum:	+ 1,721
Add:	2,992

The answer is always a palindrome!

> **Note:** *Sometimes it will take several steps. Keep reversing the sum and adding. Eventually, you will get a palindrome!*

Sharp Ears for Math

Here is a quick math exercise that can be completed without pencil and paper. Ask students these questions. Use mental math for the addition and subtraction. Remind students to use their sharpest listening skills. Let them work in pairs to come up with an answer for each question.

1. What is 5,806 increased by 161?
2. How much greater is 6,032 than 3,999?
3. What is 5,555 increased by 4,444?
4. What is 853 minus 80?
5. What number minus 25 plus 66 equals 166?
6. How much less is 80,600 than 170,000?
7. How much greater is 100,000 than 14,000?
8. What is 99 plus 3,660?
9. What number is 550 less than 5,500?
10. What number is 790 more than 1,660?

Something's Missing

Sharp addition and subtraction skills are needed here. Bring in receipts from grocery store visits, one for each student. Use these as examples of column addition. With a black pen, block out one of the numbers on each receipt. Challenge students to figure out what that number was. They will need to perform addition and subtraction to track down the missing number. (You can also make up your own columns with blocked-out numbers. Make sure the columns show the sum!)

Milk
2 lbs Apples   1.99
Kleenex   4.60
Meat   1.22
Canned Corn   12.15
Frozen Peaches   .33
Detergent   ████
Cat Food   6.19
Total   7.50
                    36.97

Some Tall Numbers

Use the tall statistics from the chart to answer the questions.

1. How much taller is Petronas Towers than the tallest fountain? _____

2. How much shorter is the sand sculpture than the Sears Tower?_____

3. How much taller is the wedding cake than the Vertigo shoes?_____

4. How much taller is the fountain than the sand sculpture?_____

5. How much shorter are the shoes than the chimney?

   _____

6. How much shorter is the cake than the fountain?

   _____

7. What is the difference between the Sears Tower and the fountain? _____

8. What is the difference between the wedding cake and the Gateway Arch? _____

	Rounded to nearest whole number
World's tallest office building Petronas Towers, Malaysia	1,483 ft (452 m)
U.S. tallest building Sears Tower	1,450 ft (443 m)
World's Tallest chimney Coal power plant, Kazakhstan	1,378 ft (420 m)
World's tallest monument Gateway Arch, USA	630 ft (192 m)
World's tallest fountain Fountain Hills, AZ, USA	562 ft (171 m)
Tallest high-heeled shoes Vertigo shoes	11 in (28 cm)
Tallest sand sculpture Holland Sand Sculpture Co.	73 ft (21 m)
Tallest wedding cake Serendipity Restaurant, NY	9 ft (3 m)

Strange But True

Tell students to choose any four numbers from 0 to 9, and arrange them to make the largest possible number. Then, tell them that you will show them a way to turn their number into this number:

6,174.

Ask them to follow these instructions:

1. Write your 4-digit number.
2. Rearrange the 4 digits to make the smallest number possible.
3. Subtract that number from the larger number.
4. Arrange the digits of that answer to make the largest possible number.
5. Arrange the digits in the answer to make the smallest possible number.
6. Subtract the smaller from the larger number.
7. Keep repeating steps 4 through 6. Eventually, you will end up with the number 6,174. This works no matter what digits you choose at the beginning.

Quirky Questions

Give students these quirky questions to answer. You can write the number on the board, then ask the question. Allot one minute for answering each question. Students can work in pairs or small groups.

A.	48,326 scorpions	How many legs? (8 legs on each scorpion)
B.	120,513 kindergartners	How many fingers and toes?
C.	96,489 chickens	How many wings?
D.	362,101 sets of triplets	How many babies?
E.	37,816 wings	How many dragonflies? (4 wings each)
F.	749,355 ladybugs	How many legs? (6 legs on each ladybug)
G.	7,200 quadruplets	How many sets?
H	604,000 lion legs	How many lions?
I.	1,012,332 eggs	How many dozen?
J.	123,456 tripods	How many legs?

Big Money

Divide students into pairs to solve these problems about high salaries for professional athletes. Read one problem at a time to students. They should listen for and record the important facts, and find the answer. When all pairs have finished a problem, let them compare answers.

1. A heavyweight boxer earns the following payments for three fights: $31 million, $19 million, and $25 million. What is his average earning per fight?

2. A basketball player has a 7-year, $50 million contract. What is his average salary per year?

3. A football team offered a $25,500,000 contract to a 300-pound linebacker. What did the team pay per pound for this player?

4. A player with a $25,200,000 one-year contract played in 180 games during that year. What did he earn per game?

5. A basketball player scored 2,000 points in one year. His salary that year was $22 million. How much did the team pay him per point?

Top Speeds

Duplicate the "Top Speeds" chart on the board where all students can see it. They can refer to the chart as they calculate quick answers to these questions:

1. What is the speed of a train that is three times as fast as a cheetah?
2. What animal is five times faster than a chicken?
3. Which animal is about twice as fast as an elephant?
4. The pig is about five times as fast as what animal?
5. What animal is twice as fast as a black mamba?
6. What animal is thirty-five times faster than a spider?
7. Which animal travels ⅓ the speed of the elk?
8. Which animals are three times as fast as a pig but not 4 times as fast?
9. Which animal is about twice as fast as a giraffe?
10. Which animal runs at twice the speed of a wild turkey?

TOP SPEEDS	
Animal	**MPH**
Cheetah	70
Antelope	61
Lion	50
Elk	45
Coyote	43
Zebra	40
Greyhound	40
Rabbit	35
Giraffe	32
Grizzly	30
Elephant	25
Black Mamba (Snake)	20
Wild Turkey	15
Pig	11
Chicken	9
Spider	2

The 3-Digit Trick

Here is a great division trick to try again and again. Lead students through it five times, asking them to choose a different number each time.

Step 1   Choose any 3-digit number.

Step 2   Repeat the digits to make a six-digit number.

Step 3   Divide by 13.

Step 4   Divide by 11.

Step 5   Divide by 7.

Sample:

526

$$526{,}526 \div 13 = 40{,}502$$

$$40{,}502 \div 11 = 3{,}682$$

$$3{,}682 \div 7 = 526$$

Here's the trick: The answer will ALWAYS be the original number!

Elusive Numbers

Track down these numbers by using the clues. See how many you can find in ten minutes!

Find a number that equals ...

_____ 1) toes on a quartet *times* days in April

_____ 2) 79 dozen *minus* 948

_____ 3) total shoes in 1,001,687 pairs

_____ 4) legs on 468 flies

_____ 5) players' ears on a basketball court *times* players' legs on a football field

_____ 6) inches in 5 yards *minus* minutes in an hour

_____ 7) 28 centuries *divided by* 8 decades

_____ 8) elbows on 46 mosquitoes *plus* feet in 4,209 miles

_____ 9) lines in a limerick *times* grams in a 6 kilograms

_____ 10) minutes in 27 hours

_____ 11) pints in 82.5 gallons

_____ 12) this year *minus* months in 3 years

Amazing Seven

This great math trick works with any number and gives good practice in all operations. Read the instructions slowly to students while they perform the math on paper. Once you are finished with the trick, try it again with other numbers.

Step 1: Choose any number with any number of digits.

Step 2: Subtract 2.

Step 3: Multiply by 3.

Step 4: Add 12.

Step 5: Divide by 3.

Step 6: Add 5.

Step 7: Subtract the original number.

Here is the trick: No matter what number is chosen, the answer will always be 7.

$$
\begin{array}{r}
35 \\
-2 \\
\hline
33 \\
\times 3 \\
\hline
99 \\
+12 \\
\hline
111 \div 3 = 37 \\
+5 \\
\hline
42 \\
-35 \\
\hline
\mathbf{7}
\end{array}
$$

Bubble Math

Have fun blowing bubbles while you practice your skills at calculating averages.

Divide the class into small groups of three or four students. Give each student a jar of bubble liquid and a bubble-blowing tool. Each student in the group takes two turns blowing bubbles. While one student blows, the other students count the bubbles and record the number. Then, they calculate the average number of bubbles per turn for their group.

Groups can combine their data and calculate an average number of bubbles per turn for the entire class.

Curious Averages

Check up on your class's average-calculating skills. See how many of these curious averages students can find in ten minutes.

- average height of students in inches
- average shoe size
- average number of siblings
- average age (include everyone in the room)
- average weight of textbooks
- average length of ears
- average circumference of head
- average number of teeth
- average distance between eyes
- average amount of money found in student pockets

Fractions Around Every Corner

Use the people in your school or classroom to practice fraction writing. Divide students into pairs and ask them to see how many of these fractions they can accurately write in ten minutes. In each case, the fraction should show a ratio that compares the answer to the instruction to the total number in the group named (such as class, school, etc.).

Write a fraction to show...

- female students (class)
- students wearing white shoes (class)
- males wearing T-shirts (class)
- messy desks (all desks in classroom)
- people in school who are teachers (whole school)
- students who rode a bike to school today (class)
- students with perfect attendance this year (class)

- portion of school day taken up by lunch period (whole school day)
- male students (class)
- students with curly hair (class)
- sharpened pencils (pencils in classroom)
- students who ate school lunch today (school)
- students wearing glasses (class)
- black shoes (all shoes in classroom)
- students new to school this year (class)

Pizza Problems

Read each problem to students. Let them work in pairs to write an answer for each question. The answer must be written in the form of a fraction or a mixed numeral.

A. pizza cut in 10 slices   Jake ate 3. Sam ate 4.   How much pizza is left?

B. pizza cut in 6 slices   2 pieces are left.   How much pizza was eaten?

C. pizza cut in 12 slices   3 friends ate 3 each.   How much pizza is left?

D. pizza cut in 8 slices   Only 2 have olives.   How much pizza is without olives?

E. pizza cut in 16 slices   2 friends ate 2 each.   How much pizza is left?

F. 7 pizzas, 10 slices each   22 slices are left.   How much pizza was eaten?

G. 3 pizzas, 8 slices each   14 slices are left.   How much pizza was eaten?

H. pizza cut in 11 slices   4 slices have onions.   How much pizza has no onions?

I. 4 pizzas, 8 slices each   17 slices were eaten.   How much pizza is left?

J. 6 pizzas, 12 slices each   4 friends ate 5 each.   How much pizza is left?

Order, Please!

Write each of the seven fractions below in large numerals on a piece of drawing or typing paper. Pin one paper on the shirt of each of seven students.

Challenge the rest of the students to put the numbers in order. Set a timer for ten minutes. Students will need to figure out how the fractions relate to each other in size and line up the seven fractions in order from least to the greatest.

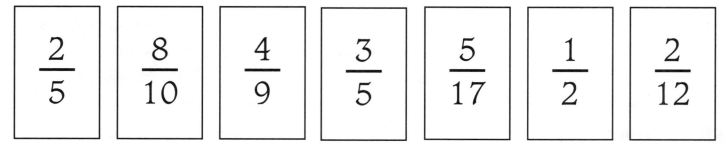

$$\frac{2}{5} \quad \frac{8}{10} \quad \frac{4}{9} \quad \frac{3}{5} \quad \frac{5}{17} \quad \frac{1}{2} \quad \frac{2}{12}$$

If students finish the task quickly, turn over the papers and write seven new fractions. Students can put these in order, too.

Find the Imposter

In each group of fractions, one is an imposter. All of the other fractions in the group are equivalent to one another. Write each group of fractions on a card. Give each card to a small group of students. Ask them to find the imposter in two minutes. Then, have the groups switch cards. Give two minutes for each search.

A. $\frac{2}{3}$ $\frac{26}{39}$ $\frac{10}{15}$ $\frac{6}{9}$ $\frac{8}{12}$ $\frac{5}{9}$

B. $\frac{9}{15}$ $\frac{21}{35}$ $\frac{7}{11}$ $\frac{6}{10}$ $\frac{30}{50}$ $\frac{18}{30}$

C. $\frac{12}{54}$ $\frac{1}{5}$ $\frac{2}{9}$ $\frac{16}{72}$ $\frac{20}{90}$ $\frac{8}{36}$

D. $\frac{6}{8}$ $\frac{21}{24}$ $\frac{49}{56}$ $\frac{28}{32}$ $\frac{35}{40}$ $\frac{77}{88}$

E. $\frac{15}{45}$ $\frac{3}{9}$ $\frac{16}{48}$ $\frac{15}{45}$ $\frac{2}{6}$ $\frac{10}{33}$

How Low Can You Go?

Try this fast-trade activity to see if your students are sharp at spotting fractions in lowest terms. Provide one piece of drawing paper to each student. Number each sheet and distribute them. Ask students to write eight fractions on their paper. They may choose any eight fractions, but only one fraction can be in lowest terms.

When the papers are ready, each students should make a numbered list, with as many numbers as there are students in the class. Ring a bell or give a signal. At the signal, students trade papers and look for the fraction in lowest terms. When they recognize it, they can write the fraction on their list (next to the number that corresponds to that paper). Sound the bell every thirty seconds for a switch of papers.

Some "Ant"-ics

Practice adding and subtracting fractions while you help an ant get to the top the hill. The ant begins at the bottom of a meter-high hill. Students need pencil and paper and their skills with fractions to follow its progress. Students may work in pairs. They should write a fraction in lowest terms to show where the ant is at the end of each hour. Read the following information to them. All calculations must be done in fractions.

Hour 1: The ant climbed $\frac{8}{10}$ meter.

Hour 2: The ant slid backwards $\frac{1}{3}$ meter.

Hour 3: The ant slid backwards $\frac{2}{10}$ meter.

Hour 4: The ant climbed uphill $\frac{2}{4}$ meter.

Hour 5: The ant slid back $\frac{5}{30}$ meter.

Hour 6: The ant climbed forward $\frac{1}{5}$ meter.

Hour 7: The ant climbed $\frac{1}{10}$ meter.

Hour 8: The ant slid back $\frac{2}{3}$ meter.

Hour 9: The ant climbed $\frac{4}{15}$ meter.

Hour 10: The ant climbed $\frac{3}{6}$ meter.

Too Much Food

What do you do to get the right amount of sweets? This yummy recipe for Supremely Sweet Snack Mix serves 36 hungry eaters. But only 24 guests are coming to tomorrow's party. Reprint the recipe for students or write it on the board. Ask students to alter the recipe so that it serves 24. (They will need to multiply each amount by ⅔.)

Supremely Sweet Snack Mix

2 ½ gallons	sugared granola
2 ⅔ quarts	dried peaches
2 bags	gumdrops
1 ⅖ cups	sweetened coconut
5 ½ cups	chocolate candies
4 cups	brownie crumbs
3 ½ bags	miniature marshmallows
3 ⅞ cups	chocolate chips
1 ¾ cups	raisins
4 ¾ cups	jellybeans

Toss all ingredients together lightly.

Serves 36

Only 3 Answers

Practice dividing fractions with fifteen problems. Each problem will result in one of these three answers. Write the three answers (in large type) on the board or on a poster. Students must group each problem under one of the three answers. See if they can finish all fifteen in ten minutes.

The answers are: $\frac{3}{4}$ $\frac{1}{2}$ $\frac{1}{3}$

The problems are:

A. $\frac{5}{10} \div \frac{2}{3}$ F. $\frac{3}{3} \div \frac{4}{3}$ K. $\frac{7}{4} \div \frac{7}{2}$

B. $\frac{4}{5} \div \frac{8}{5}$ G. $\frac{3}{11} \div \frac{9}{11}$ L. $\frac{2}{4} \div \frac{4}{6}$

C. $\frac{5}{2} \div \frac{25}{5}$ H. $\frac{5}{12} \div \frac{5}{6}$ M. $\frac{2}{3} \div \frac{16}{8}$

D. $\frac{4}{16} \div \frac{6}{8}$ I. $\frac{13}{26} \div \frac{2}{3}$ N. $\frac{5}{25} \div \frac{2}{5}$

E. $\frac{3}{7} \div \frac{9}{7}$ J. $\frac{2}{3} \div \frac{24}{12}$ O. $\frac{5}{16} \div \frac{5}{8}$

Ratios Come to School

Ratios are all about comparison. A ratio compares one amount to another with a fraction. See how many of these school ratios students can write in ten minutes.

1) 18 lessons, 4 days

2) $540 spent on a field trip, 30 students went on the field trip

3) 5 As on the test, 35 students took the test

4) 300 miles traveled to school, 120 students

5) 7 students with perfect attendance, 49 students in the class

6) 38 broken lockers, 342 locker owners

7) 462 items in the lost and found, 77 students in the school

8) 23 teachers in the school, 552 students in the school

9) 7,275 French fries eaten at lunch in the cafeteria, 485 students ate French fries for lunch

10) 2,340 pencils in the school; 468 students, teachers, and staff in the school

The Mobile Decimal Point

The slight move of a decimal point makes a major difference in a number. Write a decimal numeral on the board. Ask a student to read the number. Move the decimal point to another location. Ask another student to read the number.

Move the decimal point three times or more for each number.

EXAMPLE:

206.25	*two hundred six and twenty-five hundredths*
20.625	*twenty and six hundred twenty-five thousandths*
0.20625	*twenty thousand, six hundred twenty-five hundred thousandths*
2.0625	*two and six hundred twenty-five ten thousandths*
2,062.2	*two thousand sixty-two and two tenths*

Decimal Detectives

In each of these numbers, a digit shows up at least twice. Write the numbers on the board or on cards. Show them, one at a time, to students. The students should find the digit and write, in words, the different values of the digit.

Numbers:

A. 3.5343

B. 79.0727

C. 0.222

D. 15.9811

E. 6.67896

F. 55.50843

G. 105.0803

Examples:

440.147	The digit four is worth four hundreds, four tens, and four hundredths.
239.0996	The digit nine is worth nine ones, nine hundredths, and nine thousandths.
86.5005	The digit five is worth five tenths and five ten-thousandths.

Money Matters

Use any amount of money for a quick practice in all four operations with decimals. Students start with any amount of money and apply different operations to each amount. (For time purposes, limit the amount to $1,000 or less.)

Direct students to choose any amount of money (under $1,000) in dollars and cents, and to follow each instruction as it is given, starting with that original amount.

- Increase it by 152.50.

- Decrease it by 89.14.

- Multiply it by 37.5.

- Divide it by 15.

- Find 45% of it. (.45)

- Double it.

- Add it to $90,000.00

- Subtract it from $100,000.00

- Triple it.

- Find ⅓ of it.

No Fractions Allowed

Today is the day to ban fractions from math class. Divide students into small groups. Give each group this list of items. Tell them to get rid of the fractions by converting them all to decimals (rounded to the nearest hundredth). Challenge each group to finish as many conversions as they can in ten minutes.

1. The book has an average of 130½ words per page.

2. The wastebasket holds 17⅔ pounds of junk.

3. The average height of students is 55⅞ inches.

4. The globe is 4⅘ meters from the doorway.

5. Jon's earlobe measures 3⅓ centimeters.

6. Micky finished 44⅕ seconds behind Joe.

7. Maria's pet elephant weighs 1,788 5/12 pounds.

8. Louis ate 23 8/9 grams of pizza.

9. The teacher's wig cost 98¼ dollars.

10. Julie ate 9/12 of her lunch today.

11. Alex lost ⅝ of his allowance.

12. Mac's pet snake is 3⅗ feet long.

13. The sled dog trip took 18⅚ days.

14. The movie lasted 106 11/12 minutes.

Purchases in a Hurry

A scuba diver is on his way to an important dive when he discovers that his equipment has been stolen. Luckily, the sporting goods store has a sale on scuba gear today. Give students the challenge of helping the diver make decisions about replacing his equipment in a hurry. Use the prices and information on the sign from the store. Ask students to solve these problems:

1) What will he pay for a wet suit?
2) What amount can be saved on the gauges?
3) How much will a mask and snorkel cost?
4) What will he pay for a camera?
5) What is the savings on fins and 2 tanks?
6) What will he pay for a mask?
7) What is the savings on the tanks?
8) He has $450. Can he buy everything?

SALE ON SCUBA GEAR
TAKE 30% OFF ALL SCUBA GEAR

	Original Price
2 tanks	$320
wet suits	$168
underwater camera	$145
fins	$78
gauges	$65
masks	$49
snorkels	$22

Fast Food & Quick Thinking

Practice percentages by calculating taxes and tips. Start with your favorite lunch. Ask students to write their own ticket from a lunch of their choice. The ticket should include the price of each item ordered.

Then, students can use their personal menus to answer these questions.

1) If there is a 5% tax on the food, what will the total bill be?

2) If there is an 8% tax on the food, what will the tax be?

3) You want to leave a 10% tip. What will that be?

4) Senior citizens get a 15% discount. What would this lunch cost a senior?

5) If the tax is 3% and you leave a 15% tip on the total food bill (not including the tax), what will your total cost be?

6) You have $20. The tax is 7%. Can you afford to leave a 20% tip?

My Favorite Lunch	
taco plate	4.50
salad	2.45
milkshake	2.35
brownies	1.95
	11.25

Rates at the Race Track

The race track is a good place to think about rates: rates for the speed of vehicles, rates for the speed of services, or rates for the consumption of things such as gasoline and oil. Ask students to think about different rates that would "hang around" a racetrack. Brainstorm some ideas together. Then, group students into pairs and ask them to create ten rate problems that have to do with things happening at a racetrack.

Sample Problems

- **Gallons per hour?** Cars burned 256 gallons of fuel in 4 hours.
- **Miles per minute?** One car traveled 21 miles in 6 minutes.
- **Miles per gallon?** 15 gallons of gas used in 105 miles.
- **Quarts per car per hour?** 30 cars burned 180 quarts of oil in 2 hours.
- **Hot dogs eaten per minute?** Fans ate 8,760 hot dogs in 2 hours.

Ideas for Rates

- speed of vehicles (or horses, dogs, or people)
- consumption of gasoline
- speed of services
- consumption of food

Geo-Hunt

The everyday world is full of geometry. Set up a short "geo-hunt" in the classroom, school building, or on the school grounds. Divide students into small groups, and give this list of geometric figures to each group. Give them ten minutes to locate and record the name of an item that contains each figure.

- line segment
- plane
- parallel lines
- intersecting lines
- right angle
- straight angle
- obtuse angle
- acute angle

- isosceles triangle
- scalene triangle
- equilateral triangle
- square
- circle
- rhombus
- rectangle
- parallelogram

- trapezoid
- hexagon
- rectangular prism
- sphere
- cube
- pyramid
- cone
- cylinder

Who Am I?

Check up on understanding of the properties of figures with this riddle game. Read each riddle. Students can answer the riddle with the name of a geometric figure. Let students make up other geometry riddles.

1) I'm a 10-sided polygon. Who am I?

2) I'm a plane figure with 3 angles and 3 sides. Who am I?

3) I'm a space figure with a square base, 4 other faces, 5 vertices, and 8 edges. Who am I?

4) I'm a 3-sided polygon with no equal sides and no equal angles. Who am I?

5) I'm a plane figure with two rays that meet at a vertex. Who am I?

6) I'm a quadrilateral with only 1 pair of parallel sides. Who am I?

7) I'm a space figure with no edges and no vertices. Who am I?

8) I'm a 3-sided polygon with no right angles and no angles > 90. Who am I?

9) I'm a space figure with 2 bases, 1 other side, 2 edges, and no vertices. Who am I?

10) I'm a prism with 3 sets of square, parallel bases. Who am I?

11) I'm a parallelogram with four equal angles. What am I?

12) I'm a quadrilateral with no right angles and 4 equal sides. Who am I?

Signature Symmetry

Strengthen the understanding of symmetry by turning signatures into symmetrical creatures. Review the concept of symmetry with students. Quickly draw or find items that are symmetrical or non-symmetrical. Give each student a large piece of colored construction paper, light-colored chalk, and a pair of scissors. Read these directions to the students.

1. Fold the paper in half lengthwise.

2. Using the fold as a line, write your name in large letters. The letters should almost touch the top edge of the paper. Make sure each letter touches the line and that each letter touched the next letter. Leave the "tails" off of the letters.

3. Cut around your name, away from the lines of writing to make large, fat letters. Cut out the centers of any letters such as o, e, d, g, etc.

4. Wipe off the chalk. Open the design and glue it onto a piece of paper of a contrasting color.

Puzzling Perimeters

Calculate and compare perimeters of different figures. Write this list on the board. Students can work together in small groups to find the perimeter that answers the question for each group.

Group # 1: Greatest perimeter? A. a 6-in square
   B. a 7 x 9-cm rectangle

   C. a circle with a 10-cm diameter
   D. a triangle with three 9-in sides

Group # 2: Greatest perimeter? A. a circle with 5-m diameter
   B. a trapezoid with 4 x 6 x 5 x 7-in sides

   C. a 7-inch square
   D. a 15 x 6-m rectangle

Group # 3: Least perimeter? A. a 8 x 20-m backyard
   B. a 15-ft square dance floor

   C. a 40 x 5-m rug
   D. a 30 x 10-ft ceiling

Group # 4: Greatest perimeter? A. 19 x 5-foot chalkboard
   B. a puddle with a 3-yd diameter

   C. a 19 x 3-m window
   D. a 6 x 4-yd garden

Got You Covered!

Picture a table covered with these items. Figure out how much space each one covers. Read the descriptions to students. Give them one minute to find the area covered by each item.

1) a cookie with an 8-cm diameter

2) a 7-in square napkin

3) a 9 x 12-in casserole dish

4) a 12 x 11-cm piece of cake

5) a doughnut with a 10-cm diameter and a center hole with a 2-cm diameter

6) a 9 x 9-in baking dish

7) a 50 x 10-cm dish

8) a cup with a 4-inch diameter bottom

9) a 6-inch (diameter) pie pan

10) a 40 x 20-cm platter

11) Describe an item that would cover 16 in^2.

12) Describe an item that would cover 80 cm^2.

Hold Everything!

See how quickly students can answer this challenging volume problem. Write the dimensions of each container on the board.

The problem: The Cooking Club mixed up 6,000 cubic centimeters of a tasty Spicy Surprise Soup. Which of the following containers could hold the soup?

    A. a cylinder that is 30 cm tall with 10-cm diameter bases

    B. a rectangular prism that is 10 cm x 24 cm x 30 cm

    C. a spherical container with a 5-cm radius

    D. a pyramid with a 15-cm square base and a 30-cm height

    E. a 20 x 20 x 18 cm cube

    F. conical container with a 35-cm height and a base with a 10 cm radius

Unit Wisdom

Sharpen students' skills at choosing appropriate units of measure with this quick round-the-room challenge. Group students into pairs. Read a question to a pair of students. Ask them to consult and give a quick answer. (There is more than one correct answer for many of the questions.) Then, move on to the next pair. Try to get an answer to all the questions.

What unit of measurement would you use to find ...

1. the time it takes to run ten feet?
2. the amount of water in a bird bath?
3. the air inside a balloon?
4. the width of a button?
5. the amount of coffee in a cup?
6. the time between heartbeats?
7. the distance between two cities?
8. the length of a crayon?
9. the temperature of an iceberg?
10. the size of a continent?
11. the weight of a whale?
12. the amount of water in a raindrop?
13. the length of a trip into space?
14. the depth of a swimming pool?
15. the space inside a tent?
16. the length of a decade?
17. the weight of a monkey?
18. the distance between your eyes?
19. the area of a cookie?
20. the amount of water in an ocean?

Conversion Quest

With this quick contest, students can polish up the skill of converting measurements.

Divide the class into groups of about four students. Give this list to each group. (Photocopy it from this page, or you might read each item, allowing thirty seconds for each question.) Challenge groups to answer all twenty questions in ten minutes.

1) 15 m	How many cm?	11) 256 C	How many gal?
2) 3 L	How many mL?	12) 20 yds	How many ft?
3) 2 Tons	How many lbs?	13) 30 hrs	How many secs?
4) 10.5 km	How many m?	14) 18.3 mL	How many L?
5) 22 gal	How many qts?	15) 64 oz	How many lbs?
6) 5,000 mg	How many g?	16) 7 mi	How many ft?
7) 47 decades	How many yrs?	17) 15 yd	How many in?
8) 70 Kg	How many g?	18) 210 min	How many hrs?
9) 35 gal	How many pts?	19) 5,000 cm	How many m?
10) 16 T	How many tsp?	20) 160 lbs	How many oz?

30-Second Measurements

Try to capture these measurements in thirty seconds each. Give the following list to students. Let them work in pairs or small groups to find something satisfying each description. Let them search for ten minutes. They will need rulers and measuring sticks with both U.S. customary and metric units. Items they find may be exactly or close to the measurements listed.

Find something that...

is 1 foot long

is 3 centimeters wide

is < 6 inches long

is 2 inches thick

is < 30 centimeters long

is ½ inch wide

is 1 meter long

is 2 yards tall

is > 2 ½ feet tall

is 5 centimeters long

is 10 centimeters thick

is 2 meters tall

is 50 centimeters high

is 5 inches tall

is < 2 centimeters tall

is > 10 meters long

has a 10-inch diameter

has a 30-centimeter diameter

has a 40-inch perimeter

has a perimeter < 6 centimeters

Massive Questions

Mass is the amount of matter in an object. One way to measure mass is to weigh an object. Do some quick thinking to decide which object weighs more in each of these pairs. Make this a mental activity. Discuss the following questions as a group and make a decision for each one. If time permits, students can calculate the precise amounts by converting both items into the same unit of measure. Students can enjoy making up "massive" questions of their own.

Which weighs more...

1. a) a ½ ton school bus   or   b) a 2,600 pound elephant?
2. a) a 5 gram creature   or   b) an 8 ounce creature?
3. a) a 2 pound notebook   or   b) a 1 kilogram boot?
4. a) a 1 ounce feather   or   b) a 1 gram cotton ball
5. a) a 40 pound dog   or   b) a 30 kilogram dog?
6. a) a 1,000 milligram drop   or   b) a 2 gram drop?
7. a) a 10 ounce bug   or   b) a 100 milligram bug?
8. a) a 100 pound gorilla   or   b) 150 kilogram gorilla?

Moon Weights

Because of the difference in gravity between the moon's surface and Earth's surface, weights on Earth are about six times weights on the moon. Figure out the moon weight of each of these:

	Earth weight	Moon weight		Earth weight	Moon weight
1) an asteroid	252 lb	_____	7) a moon rock	5.5 lb	_____
2) a space ship	3 T	_____	8) unidentified creature	36.6 kg	_____
3) an astronaut	180 lb	_____	9) space lunch	504 g	_____
4) a space boot	1 kg	_____			
5) an astronaut	77 kg	_____			
6) space monkey	14 kg	_____			

10) What is the difference between the moon weights of two astronauts who weigh 60 kg and 78 kg on Earth? _____

Timely Problems

This activity takes sharp listening and quick thinking. Students may use pencil and paper to do their calculations as you read the questions. If time permits, ask students to create their own timely problems and trade them with one another for solutions.

1) It is as many hours before 1 PM as it is after 1 AM. What time is it?

2) Tom slept 14 hours and got up 5½ hours before noon. What time did he go to bed?

3) Maxine traveled for 70 hours. She arrived at her vacation spot on Friday at 11 AM. What time and day did she leave home?

4) Rob got home from his trip Monday at 1:30 PM. The trip took 16½ hours. What time did he start the trip?

5) Andy got up at 9 AM after sleeping 11 hours. Randy went to bed 2 hours before Andy. What time did Randy go to bed?

6) Abby slept 15¼ hours and got up 3 hours before midnight. What time did she go to bed?

10-Minute Tallies

Gathering statistical data begins with identifying the data needed and counting it. But the data won't do any good unless there is a way to record it. A tally sheet is a quick way to keep track of data. Students can practice gathering data with quick counts and tallies. Divide students into small groups. Give each group two or three categories of data such as those below. They should use a tally method to find the statistics within the class. Students may add other categories to the list.

- colors of shoes on classmates
- most watched TV shows
- years individuals have attended this school
- years experience of teachers in the school
- sizes of students' families
- number of freckles per student
- numbers and kinds of pets

The Three Ms

Mean, median, and mode are three key terms in statistics. Review the meaning of each term with students. Divide students into groups of four or five so they can collect data quickly. Give each group a copy of the table below. First, each group should complete the second column of the table. Then, they can find the mean, median, and mode for each set of statistics. *(See sample item done.)*

Description	List each Statistic	Mean	Median	Mode
Height of group members (in inches)	55, 53, 55, 52, 50	53	53	55
Age of each group members				
# of hours spent watching TV yesterday				
# of hours spent doing homework yesterday				
# of hours slept most nights				

24 Hours at a Glance

"What have you been doing the last 24 hours?" Ask students this question. They can use paper and pencil to do a quick review of their activities for the past 24 hours and summarize it into statistical data. (They can make lists of activities accompanied by the amount of time, in hours, spent on each activity.) Give each student a large cardboard circle (8-inch or 20-centimeter diameter or larger). With a ruler, pencil, and crayons or markers, they should create a circle graph to display their 24 hours of activity. Glue the graphs to a contrasting background where the graphs can be titled and labeled with any necessary key information.

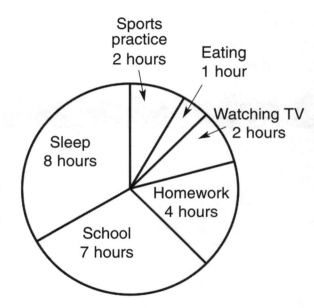

Strange Statistics

Collection	Record Number Collected
mousetraps	2,334
golf balls	48,824
shoes	10,000
light bulbs	60,000
airsickness bags	2,112
parking meters	292
ballpoint pens	108,500
underwear items	1,700
jet fighters	100
nutcrackers	2,220
gnomes & pixies	2,010

Collections of statistical data are meant to be used. They can be used to answer questions or solve problems. This table holds some strange statistics about curious collections. (These are actual records held for collecting.) Provide students with a copy of the table. Give them 10 minutes to create 15 questions that someone could answer by using the data provided on the table.

Sample Questions

What's the difference between the largest and smallest collections?

Which collection is about 5 times the size of the shoe collection?

What's the Outcome?

Review the meaning of the outcome in a probability experiment. Decide on a visual method to show outcomes (a list or chart). Then, work together to identify all the outcomes for each of these probability experiments.

1) There are three flavors of ice cream: peanut butter, vanilla, and fudge. Two scoops, chosen at random, are put on a cone. What are the possible outcomes?

2) Flip a quarter two times. What are the possible outcomes?

3) Toss two dice. What are the possible outcomes?

4) A spinner has 4 equal sections of these colors: red, green, yellow, and blue. Spin the spinner twice. What are the possible outcomes?

5) A box has 5 pool balls: 1 black, 1 white, and 3 red. Choose two without looking. What are the possible outcomes?

2-Scoop Ice Cream Cone	
Flavors:	
P = peanut butter	
V = vanilla	
F = fudge	
Possible Outcomes: (6)	
PV (same as VP)	PP
PF (same as FP)	VV
VF (same as FV)	FF

Grab Bag Probability

Gather miscellaneous items in a large grocery bag. Write the contents on the outside (as shown). Then, have fun making up probability problems. Students should write a fraction showing the probability of each event. Use the information below as an example.

1. P (S) = $\frac{8}{30}$ or $\frac{4}{15}$

2. P (M) = $\frac{3}{30}$ or $\frac{1}{10}$

3. P (R) = $\frac{4}{30}$ or $\frac{2}{15}$

4. P (T) = $\frac{10}{30}$ or $\frac{1}{3}$

5. P (B) = $\frac{5}{30}$ or $\frac{1}{6}$

6. P (S or B) = $\frac{13}{30}$

7. P (not R) = $\frac{26}{30}$ or $\frac{13}{15}$

8. P (not T) = $\frac{20}{30}$ or $\frac{2}{3}$

9. P (R or T) = $\frac{14}{30}$ or $\frac{7}{15}$

10. P (T or S) = $\frac{18}{30}$ or $\frac{3}{5}$

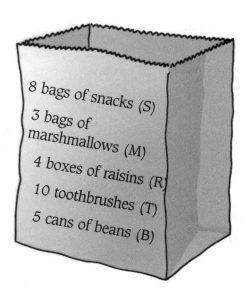

8 bags of snacks (S)

3 bags of marshmallows (M)

4 boxes of raisins (R)

10 toothbrushes (T)

5 cans of beans (B)

Double Spin

Draw the following "spinners" on the board or on a large chart. Draw the chart that shows the possible outcomes that could result from spinning each spinner once. Ask students to use the chart and spinners to answer the following questions. What is the probability of...

1. 4 and green?   P (4,g) = _____

2. 2 and blue?   P (2,b) = _____

3. 8 and green?   P (8, g) =_____

4. < 8 and purple?   P (<8, p) = _____

5. 4 and not blue?   P (4, not b) = _____

6. 2 and not green?   P (2, not g) = _____

7. > 2 and blue?   P (>2, b) = _____

8. > 2 and not purple?   P (>2, not p) =_____

9. 2 or 4 and blue?   P (2 or 4, b) = _____

10. 4 or 8 and green?   P (4 or 8, g) = _____

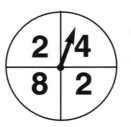
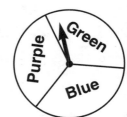

Outcomes for Spinning Both Spinners at Once			
2, p	4, p	2, p	8, p
2, g	4, g	2, g	8, g
2, b	4, b	2, b	8, b

Integer IQ

Operations with integers can be tricky to say the least. Two (or three) answers are offered for each of these problems. Have students work in pairs to examine each integer problem and select the correct answer.

1) (-14) + (36)   A. 22   B. –50   6) (-100) x (-6)   A. –600   B. 600

2) (-22) – (9)   A. –13   B. –31   7) (-51) ÷ (-3)   A. 17   B. –17

3) (88) + (-7)   A. 95   B. 81   8) (180) ÷ (-6)   A. –30   B. 30

4) (16) – (-5)   A. 11   B. 21   9) (215) + (-70)   A. –145   B. 145

5) (43) x (-9)   A. –387 B. 387   10) (-6) – (-43)   A. –37   B. –49   C. 37

How Do You Say It?

An expression is a way to "say" with mathematical symbols and numerals what can also be said with words. Read the word expressions slowly to students. Ask them to "say" the same thing with numerals and symbols (write a mathematical expression).

1) three times a number **x**

2) fifteen less than a number **y**

3) a number **z** divided by negative five

4) fifty multiplied by a number **d**

5) thirty divided by the sum of seven and a number **k**

6) negative ten multiplied by the difference between a number **w** and five

7) a number **n** decreased by one hundred

8) fifteen times the product of seven and a number **p**

9) triple the sum of a number **t** and forty

10) 12 times the square of a number **g**

Keep It Simple

Write these expressions on the board, one at a time. Give students 30 seconds to rewrite each expression in its simplest form. See if they can do all ten correctly within ten minutes!

1) $t + (t + 5) + (t - 75) + (t - 3)$

2) $2y + 7y < 14z + 3z - 6z$

3) $5d(2 + 3) - 13d + 3 - 9$

4) $12 b + -8 b + b(4 + 5)$

5) $100x - (-y + y) - 2x$

6) $4(b + c) - 2c$

7) $200 (100y + 3) + y$

8) $40 + 8(a + b) + a - b$

9) $5.5n - g + 3.7g$

10) $\dfrac{x + 3x + x}{5}$

Missing Marshmallows

Find the right equations to solve the marshmallow problems. Write the following six equations in large print on a chart or chalkboard where students can see them clearly. Read each problem slowly (twice). Students choose the correct equation to solve the problem.

Problem # 1: Three friends dropped a total of 35 marshmallows into the campfire.
Bo dropped 3 marshmallows in the fire.
Dan dropped 4 times as many as Bo and Charlie together. How many did Charlie drop?

A. $m = 35 + (35 - 4) + 2(35 + 35 - 4)$

B. $3 + m + 4(3 + m) = 35$

C. $4 + 2(4) + m = 35$

D. $35 + (35 - 4) + 2(35 - 4) = m$

E. $m = 35 - 4 + 2(4)$

F. $4(3 + m) = 35$

Problem # 2: Sam ate 35 marshmallows on Monday, 4 less on Tuesday, and twice as many on Wednesday as on Monday and Tuesday combined. How many did Sam eat in the three days?

Problem # 3: Squirrels carried off 35 marshmallows in 3 nights. The first night, they carried off 4. The next night they carried off twice as many. How many did they take away the third night?

Find the Word

When students solve these equations correctly, they'll find eight letters that make up a mystery word. Write the codes on the board. As students solve each equation, they find the matching letter on the code chart. When they have all eight letters, they must unscramble the letters to find the mystery word.

1) $\frac{1}{3} t = 333$

2) $3z + 22 - z = 162$

3) $2(n + 2) - 8 = -28$

4) $41 = c + 2c - 100$

5) $2r + 5 = -21$

6) $9 + 13y = 48$

7) $-7r = 84$

8) $x^2 - 40 = 60$

Codes	
10 = P	100 = Y
−10 = D	111 = R
47 = L	−12 = Z
12 = A	999 = I
70 = U	3 = N
13 = T	−13 = G

No Equations in Sight

None of today's challenges have any equal signs! These are non-equations or inequalities. Inequalities have solutions, too. Write the following numerals in large print on the board. Then, read the following inequalities to students, one at a time. Have someone circle each numeral that is a solution for the inequality. (They will need to erase the circles before moving on to the next problem.)

1) $x \geq 4$

2) $x \leq -1$

3) $x > -8$

4) $x + 4 < 9$

5) $4x < -12$

6) $\frac{1}{2} x < 2$

7) $3x < -9$

8) $x(4 + 3) \leq 7$

4	−1	−3	5	−10
3	0	−2	−6	2
−4	7	1	−12	−9

Something's Missing

A problem can't be solved unless there is enough information to solve it. Students need to be sharp at reading or hearing a problem and identifying the information needed for a solution. Read each one of these problems twice. Students can consult with one another to determine what missing information is needed for the solution.

1. A monkey ate 36 fewer bananas than his brother. How many did he eat?

2. Six of the seven ice skating judges gave Marika the same score. The total of all seven scores was 40.2. What was the seventh judge's score?

3. Surfer Cindy caught three times as many waves as surfer Mindy. How many waves did Mindy catch?

4. 5,800 fans arrived to watch the football game. How many fans stayed for the whole game?

5. 87 balloons finished the race. How many dropped out before finishing?

6. The awards banquet ended at 10:15 PM. How long did the banquet last?

Smart Starters — Math   81

Personal Problems

Ask students to collect a list of "personal" numbers. These can be ages, addresses, phone numbers, weights, amounts of money in pocket or savings, distance from home to school, waist measurement, and such. Try to do this in two or three minutes.

After the numbers are recorded, give each student four minutes to write at least one math problem that uses some of the numbers. After four minutes, trade papers so students can solve the problems created by their classmates.

<table>
<tr>
<td>Example: My address has 4 digits. All of them are under 8. Three digits are the same. The largest digit is last. The sum of the digits is 16. What is my address?</td>
<td>Example: My birth date is 11–14–1992. My age is 12. My phone number is 541–488–1234. Is the sum of all the digits in my phone number greater than the product of my birthday and my age?</td>
</tr>
</table>

Who's Not Home?

Use the following information and clues to figure out who is not at home. It will be helpful to draw a diagram to help with solving the problem.

Four neighbors with different professions live in apartments 5, 6, 7, and 8. One of them is not home. Who is it? (Give their name.)

Clues:

1. The magician is at home in Apartment 5.

2. Max lives between the plumber and the baker.

3. Samantha is not a baker.

4. George lives in Apartment 8.

5. The plumber is at home.

6. The magician does not live next to the zookeeper.

7. Zelda lives in an odd-numbered apartment.

8. The baker is at home.

Math for the Ages

Use the strategy of trial and error to find quick answers to these age problems. If students have any time left, they can make up more problems like these.

1 Five years ago, Abby was 5 times the age of her brother Bill. In one year, Abby's age will be double the age of her brother. How old is Bill?

2 Charlie is ½ the age of his brother Zach. In 4 years, Charlie will be ⅔ his brother's age. How old is Zach?

3 Nan is two years older than Stan. In 3 years, Stan will be ⅘ of Nan's age. How old are they now?

4 Mac is 3 years younger than Gary. Sam is 10 years older than Mac. The sum of their ages is 31. How old is Gary?

Use Your Head!

Get some mental exercise with these problems. Read each problem and wait about one minute for students to solve it. They may "put their heads together" and work in pairs to find the solutions.

1. Bob does 81 sit-ups a day. Todd does ⅔ as many. How many is that?

2. Angie ends her 75 minute workout at 1:10 PM. When did she start?

3. Jake's pulse climbs 110 beats per minute to 190 bpm while he exercises. What was his heartrate when he started?

4. Lulu jogs 20 times around a 12 x 10 meter mat. How far does she jog?

5. Maria's first game was Monday, June 25. What day is her last game (August 1, same year)?

6. Joe exercised 55 minutes a day on Monday and Tuesday, and 40 minutes a day every other day last week. What was his total exercise time?

7. Tomas does 95 sit-ups in 2 minutes. At this rate, how many would he do in 10 minutes?

8. Bottles of water cost $0.85 each. How many can Pat buy for $20?

What's Next?

Work together to find a pattern for each set of numbers or information. Then, use the pattern to solve the problem or fill in the missing numbers. Write the patterns on the board so students can examine them as they try to solve the problems.

1.	250	245	235	220	200	175	☐
2.	1,000	850	900	750	800	☐	☐
3.	12	36	108	324	☐	☐	8,748
4.	1,000	900	700	400	☐	–500	☐
5.	30	23	31	24	33	26	☐

6. Week #1, Julie walked 3 miles. The 2nd week, she walked 10% more. The 3rd week, she walked 20% more than the 2nd week. The 4th week, she walked 30% more than the 3rd week. Following this pattern, what will her rate be at the end of the 5th week?

Draw & Solve

Present these problems to students. Ask them to draw a picture or diagram to solve each problem.

1. Tori is behind Tracy. Trisha is ahead of Tracy. Trisha is ahead of Tori. Tori is the 4th runner behind Tami. Who won the race?

2. A group of gymnasts are standing evenly spaced around a circular trampoline. The gymnasts have consecutive numbers beginning with #1. Gymnast #14 is directly across from gymnast #6. How many are in the group?

3. Abigail walks along a 27-foot long balance beam. Each step takes her forward 2.3 feet, followed by a backwards step of .5 feet. How many forward steps will it take her to get to the end of the balance beam?

4. Evan is taller than Brad. Josh is taller than Wade and Evan, but shorter than Matt. Tom and Brad are between Josh and Matt. Who is tallest of the entire group?

Travel Time

Review North American time zones. If possible, have a map of North America in front of students as they tackle these time zone problems. These problems can be done mentally or with the help of pencil and paper.

1. Al left Chicago at 1:20 PM (CST) on Monday and drove to New York City, arriving at 12:15 AM (EST) Tuesday. How long (in time) was his trip?

2. Wally flew from Washington, D.C. to San Francisco, changing planes once. He left D.C. at noon (EST) and arrived in San Francisco 8½ hours later. What time (PST) did he arrive?

3. The time in Honolulu, Hawaii is 6 hours earlier than in Miami, Florida. Greg arrived in Miami from Hawaii at 8:35 AM Miami time (EST) on Friday after a 13 hour and 15 minute trip. What time (Hawaii time) did he leave Honolulu?

4. Dave drove from Denver to Philadelphia on a 40-hour trip. He arrived in Philadelphia at 9:15 PM (EST) on Sunday. What time (MST) and day did he leave Denver?

5. Zelda's trip from Tennessee (CST) to Seattle (PST) took 6½ hours. She left Tennessee at 11:00 pm on Wednesday. When did she arrive in Seattle?

Train Problems

Those problems about speeding trains are always fun. They are a great way to polish skills with rate-time-distance calculations. Make sure students understand the basic formula (rate x time = distance) and how to change it to find rate or time. Put the formula to work on these train problems. If time permits, ask students to create their own train problems.

1. A train travels 200 miles in 20 hours. What's the rate (speed)?

2. A train travels for 14.5 hours at 80 mph. How far does the train travel?

3. A train covers 900 miles in distance moving at 75 miles per hour. How long did this take?

4. A train travels at 85 mph for 11 hours. How far will it travel?

5. A train covers 1,620 miles in 27 hours. What is its rate (speed)?

6. Train A and Train B are 1,200 miles apart on parallel tracks. They both leave at noon and travel at 60 mph. What time will they meet?

Call the Doctor

To solve some problems, it is necessary to set up a proportion. Listen to these problems and work together to set up the correct proportion to find the solution. Then, solve the problems.

1. Of 20 skateboarders, 16 were injured. There are 80 skateboarders in the group. At this rate, how many injuries were there?

2. Two of every seven skiers left the team due to injuries. If 28 left, how many skiers started the season?

3. The football team members spent $900 on hospital bills for the first 2 weeks of the season. At this rate, how much will be spent over a 12 week season?

4. Of every nine hikers, seven got nosebleeds. There were 459 hikers. How many got nosebleeds?

5. 350 out of 500 soccer injuries were ankle injuries. Of each 10 injuries, how many are ankle injuries?

6. The ratio of climbs to falls for rock climber Tom is 4 falls for every 15 climbs. At this rate, how many times will he NOT fall in 180 climbs?

Answers

PAGE 9
Find answers in your math text glossary.

PAGE 10
1. C
2. J
3. A
4. H
5. B
6. D
7. G
8. F
9. I
10. E

PAGE 11
1. –3; 0.25; 2.05; 5.09; 5.9; 509; 5,579; 5,795; 60,101; 61,000
2. –99; –9.9; –0.9; 9.09; 9.99; 99.9; 99.99; 909; 99,099; 909,009
3. –4,321; 0.4;

⅔; 1.03; 1.2034; 12.3; 12½; 14.32; 1,234; 1,324
4. 88.8; 89.08; 99.08; 989; 998; 8,188; 8,888; 8,888.8; 8,988; 8,989
5. -83; –51.05; –23; –16; –½; 0.016; 3.65; 36; 813; 831

PAGE 12
Answers will vary.

PAGE 13
1. yes
2. yes
3. no
4. yes
5. yes
6. yes
7. yes
8. yes

PAGE 14
There may be more than one correct

answer for each.
1. 7780
2. 3431
3. 591
4. 595

PAGE 15
Answers will vary.

PAGE 16
A. 123,460; 123,500; 123,000; 120,000; 100,000
B. 887,650; 887,700; 888,000; 890,000; 900,000
C. 585,860; 585,900, 586,000; 590,000; 600,000
D. 260,970; 261,000; 261,000; 260,000; 300,000

E. 797,980; 798,000; 798,000; 800,000; 800,000
F. 483,110; 483,100; 483,000; 480,000; 500,000
G. 101,100; 101,100; 101,000; 100,000; 100,000
H. 666,670; 666,700; 667,000; 670,000; 700,000
I. 709,260; 709,300; 709,000; 710,000; 700,000
J. 347,920; 347,900; 348,000; 350,000; 300,000

PAGE 17
A. Answers will vary.
B. Answers will vary.
C. USA: MDCCLXXVI
D. MMDCXCIV
E. Answers will vary.
F. Answers will vary.
G. MCCXCV
H. MDCIL

PAGE 18
1. 3.6×10^7
2. 4.9×10^7
3. 7.5×10^4
4. 8.87×10^7
5. 1.9×10^9
6. 3.1×10^4
7. 3.0×10^7
8. 1.3×10^5
9. 2.0×10^{11}
10. 1.0×10^5

PAGE 19
A. equal

B. equal
C. equal
D. not equal
E. not equal
F. not equal
G. equal
H. not equal
I. equal
J. equal
K. not equal
L. equal

PAGE 20
1. $396.00
2. $3,978.75
3. $38,330.00
4. $81,862.50
5. $6,970.00
6. $20,750.00
7. $712.50
8. $3,906.00
   Crafty Kate has the most.

PAGE 21
Answers will vary.

PAGE 22
Answers will vary.

Answers

PAGE 23
Numbers with 2 factors: 7, 11
Numbers with 4 factors: 27, 35, 49, 121
Numbers with 6 factors: 12, 18, 20, 28, 50, 99
Numbers with 8 or more factors: 60, 66, 72, 1,000

PAGE 24
A. 1,000
B. 10,000
C. 1,000,000
D. 100
E. 10
F. 100
G. 10
H. 100
I. 10,000
J. 10
K. 100,000
L. 10

PAGE 25
A. no
B. no
C. no
D. yes
E. yes
F. yes
G. yes
H. no
I. no
J. no
K. no
L. no

PAGE 26
A. identity
B. associative
C. zero
D. associative
E. distributive
F. commutative
G. identity
H. commutative

PAGE 27
Answer is always a palindrome.

PAGE 28
1. 5,967
2. 2,033
3. 9,999
4. 773
5. 125
6. 89,400
7. 86,000
8. 3,759
9. 4,950
10. 2,450

PAGE 29
Answers will vary.

PAGE 30
1. 921 ft or 281 m
2. 1377 ft or 422 m
3. 8 ft, 1 inch or 2 m, 72 cm, or 272 cm
4. 489 ft or 150 m
5. 1,377 ft, 1 in or 419 m, 72 cm
6. 553 ft or 168 m
7. 888 ft or 272 m
8. 621 ft or 189 m

PAGE 31
All answers: 6,174.

PAGE 32
A. 386,608
B. 2,410,260
C. 192,978
D. 1,086,303
E. 9,454
F. 4,496,130
G. 1,800
H. 151,000
I. 84,361
J. 370,368

PAGE 33
1. $25 million
2. $7,142,857.14
3. $85,000
4. $140,000
5. $11,000

PAGE 34
1. 210 mph
2. elk
3. lion
4. spider
5. zebra, greyhound
6. cheetah
7. turkey
8. rabbit, zebra, greyhound, coyote
9. antelope
10. grizzly

PAGE 35
The answer will be the same as the original number.

PAGE 36
1. 1,200
2. 0
3. 2,003,374
4. 2,808
5. 220
6. 120
7. 35
8. 22,223,520
9. 30,000
10. 1,620
11. 660
12. Answers will vary

PAGE 37
The answer will always be 7.

PAGE 38
Answers will vary.

PAGE 39
Answers will vary.

PAGE 40
Answers will vary.

page 41
A. $^3/_{10}$
B. $^4/_6$
C. $^3/_{12}$
D. $^6/_8$
E. $^{12}/_{16}$
F. $^{48}/_{70}$
G. $^{10}/_{24}$
H. $^7/_{11}$
I. $^{15}/_{32}$
J. $^{52}/_{72}$

PAGE 42
Order of fractions is:
$^2/_{12}$; $^5/_{17}$; $^2/_5$; $^4/_9$; $^1/_2$; $^3/_5$; $^8/_{10}$

PAGE 43
A. $\frac{5}{9}$
B. $\frac{7}{11}$
C. $\frac{1}{5}$
D. $\frac{6}{8}$
E. $\frac{10}{33}$

PAGE 44
Answers will vary.

PAGE 45
Convert all fractions to 60ths. Answers may be equivalent fractions
Hour 1: $\frac{48}{60}$
Hour 2: $\frac{28}{60}$
Hour 3: $\frac{16}{60}$
Hour 4: $\frac{46}{60}$
Hour 5: $\frac{36}{60}$
Hour 6: $\frac{48}{60}$
Hour 7: $\frac{54}{60}$
Hour 8: $\frac{14}{60}$
Hour 9: $\frac{30}{60}$
Hour 10: $\frac{60}{60}$ or 1 m—top of the hill!

PAGE 46
1 $\frac{2}{3}$ gallons granola
1 $\frac{7}{8}$ quarts peaches
1 $\frac{1}{3}$ bag gumdrops
$\frac{14}{15}$ cups coconut
3$\frac{2}{3}$ cups chocolate candies
2$\frac{2}{3}$ cups brownies
2$\frac{1}{3}$ bags marshmallows
2 $\frac{7}{12}$cups chocolate chips
1 $\frac{1}{6}$ cups raisins
3$\frac{1}{6}$ cups jellybeans

PAGE 47
A. $\frac{3}{4}$
B. $\frac{1}{2}$
C. $\frac{1}{2}$
D. $\frac{1}{3}$
E. $\frac{1}{3}$
F. $\frac{3}{4}$
G. $\frac{1}{3}$
H. $\frac{1}{2}$
I. $\frac{3}{4}$
J. $\frac{1}{3}$
K. $\frac{1}{2}$
L. $\frac{3}{4}$
M. $\frac{1}{3}$
N. $\frac{1}{2}$
O. $\frac{1}{2}$

PAGE 48
1. $\frac{18}{4}$ or $\frac{9}{2}$
2. $\frac{540}{30}$ or $\frac{54}{3}$
3. $\frac{5}{35}$ or $\frac{1}{7}$
4. $\frac{300}{120}$ or $\frac{5}{2}$
5. $\frac{7}{49}$ or $\frac{1}{7}$
6. $\frac{38}{342}$ or $\frac{19}{171}$ or $\frac{1}{9}$
7. $\frac{462}{77}$ or $\frac{6}{1}$
8. $\frac{23}{552}$ or $\frac{1}{24}$
9. $\frac{7275}{485}$ or $\frac{15}{1}$
10. $\frac{2340}{468}$ or $\frac{5}{1}$

PAGE 49
Answers will vary.

PAGE 50
A. 3 ones, 3 hundredths, and 3 ten thousandths.
B. 7 tens, 7 hundredths, and 7 ten thousandths.
C. 2 tenths, 2 hundredths, and 2 thousandths.
D. 1 ten, 1 one thousandth, and 1 ten thousandth.
E. 6 ones, 6 tenths and 6 one hundred thousandths.
F. 5 tens, 5 ones, and 5 tenths.
G. 0 tens, 0 tenths, and 0 thousandths.

PAGE 51
Answers will vary.

PAGE 52
1. 130.5 words
2. 17.67 lbs
3. 55.88 in
4. 4.8 m
5. 3.33 cm
6. 44.2 sec
7. 1,788.42 lbs
8. 23.89 g
9. 98.25
10. 0.75
11. 0.63
12. 3.6 ft
13. 18.83 days
14. 106.92 min

PAGE 53
1. $117.60
2. $19.50
3. $49.70
4. $101.50
5. $119.40
6. $34.30
7. $96
8. no

PAGE 54
Answers may vary slightly.
Example:
1. $11.81
2. $0.90
3. $1.13
4. $9.56
5. $13.28
6. yes

PAGE 55
Problems will vary.

PAGE 56
Answers will vary.

PAGE 57
1. decagon
2. triangle
3. pyramid
4. scalene triangle
5. angle
6. trapezoid
7. sphere
8. isosceles triangle
9. cylinder
10. cube
11. rectangle
12. rhombus or parallelogram

PAGE 58
Answers will vary.

Answers

PAGE 59
#1 D
#2 D
#3 C
#4 C

PAGE 60
1. 50.24 cm²
2. 49 in²
3. 108 in²
4. 132 cm²
5. 75.36 cm²
6. 81 in²
7. 500 cm²
8. 12.56 in²
9. 28.26 in²
10. 800 cm²
11. Answers will vary.
12. Answers will vary.

PAGE 61
A. no
B. yes
C. no
D. yes
E. yes
F. no

PAGE 62
Answers may vary somewhat.
1. sec
2. L, C, pt, qt, gal, in³, cm³
3. cm³ in³, m³, ft³
4. cm, in, mm
5. C, tsp, mL, oz, g, T
6. sec
7. mi or km
8. in or cm
9. degrees
10. km² or mi²
11. tons or metric tons
12. mL, oz, g
13. mi, km
14. ft, yd, m
15. m³, ft³, yd³
16. yrs
17. kg or lb
18. mm, cm, in
19. in², cm²
20. mi³, km³

PAGE 63
1. 1,500 cm
2. 3,000 mL
3. 4,000 lb
4. 10,500 m
5. 88 qt
6. 5 g
7. 470 yrs
8. 70,000 g
9. 280 pts
10. 48 tsp
11. 16 gal
12. 60 ft
13. 108,000 sec
14. 0.0183 L
15. 4 lb
16. 36,960 ft
17. 540 in
18. 3.5 hr
19. 50 m
20. 2,560 oz

PAGE 64
Answers will vary.

PAGE 65
1. b
2. b
3. b
4. a
5. b
6. b
7. a
8. b

PAGE 66
1. 42 lb
2. 0.5 T
3. 30 lb
4. 0.166 kg or 166 g
5. 12.83 kg
6. 2.33 kg
7. 0.92 lb
8. 6.1 kg
9. 84 g
10. 3 kg

PAGE 67
1. 7:00 AM
2. 4:30 PM
3. Tuesday, 1:00 PM
4. Sunday, 9 PM
5. 8 PM
6. 5:45 AM

PAGE 68
Answers will vary.

PAGE 69
Answers will vary.

PAGE 70
Answers will vary.

PAGE 71
Answers will vary.

PAGE 72
1. pv, pf, pp, vv, vf, ff
2. hh, tt, ht, th
3. 1,1; 1,2; 1;3, 1;4, 1.5;.1,6; 2,2; 2,3; 2,4; 2,5; 2,6; 3,3; 3,4; 3,5; 3,6; 4,4; 4,5; 4,6; 5,5; 5,6; 6,6
4. rr; rg; ry; rb; gg; gr; gy; gb; yy; yr; yg; yb; bb; br; by; bg
5. bw, br, rw, rr

PAGE 73
Questions will vary.

PAGE 74
1. $\frac{1}{12}$
2. $\frac{2}{12}$
3. $\frac{1}{12}$
4. $\frac{3}{12}$
5. $\frac{2}{12}$
6. $\frac{4}{12}$
7. $\frac{2}{12}$
8. $\frac{4}{12}$
9. $\frac{3}{12}$
10. $\frac{2}{12}$

PAGE 75
1. A
2. B
3. B
4. B
5. A
6. B
7. A
8. A
9. B
10. C

PAGE 76
1. 3x
2. y–15
3. $\frac{3}{4}$–5
4. 50d

5. $\frac{30}{7} + k$
6. $-10 (w-5)$
7. $n - 100$
8. $15 (7p)$
9. $3(t + 40)$
10. $12g^2$

PAGE 77
1. $4t - 73$
2. $9y < 11z$
3. $12d - 6$
4. $13b$
5. $102x$
6. $4b + 2c$
7. $20{,}001y + 600$
8. $40 + 9a + 7b$
9. $5.5n + 2.7g$
10. $\frac{3x}{5}$

PAGE 78
1 B
2 A
3 C

PAGE 79
1. 999 (I)
2. 70 (U)
3. −12 (Z)
4. 47 (L)

5. −13 (G)
6. 3 (N)
7. −12 (Z)
8. 10 (P)
The word is:
PUZZLING

PAGE 80
1. 4,5,7
2. −1, −10, −2, −6, −4, −12, −3, −9
3. 4, −1, −3, 5, 3, 0, −2, 2, −4, 7, 1, −6
4. 4, −1, −3, −10, 3, 0, −2, −6, 2, −4, 1, −12, 5, −9
5. −10, −6, −4, −12, −9
6. −1, −3, −10, 3, −2, −6, 2, −4, 1, −12, 0, −9
7. −10, −6, −4, −12, −9
8. −1, −3, −10, 0, −2, −6, −4, 1,

−12, −9

PAGE 81
1. number of bananas eaten by the brother
2. score given by each of the 6 judges
3. number of waves caught by Cindy
4. number of fans who left
5. number of balloons that started the race
6. time the banquet started

PAGE 82
Problems will vary.

PAGE 83
Max, the zookeeper (Apt. #7) is not home.

PAGE 84
1. 7
2. 8

3. Stan is 5; Nan is 7.
4. 9

PAGE 85
1. 54
2. 11:55 AM
3. 80 beats per minute
4. 880 m
5. Wednesday
6. 310 minutes or 5 hours, 10 minutes
7. 475
8. 23

PAGE 86
1. 145
2. 650, 700
3. 972, 2916
4. 0, −1100
5. 36
6. 7.2 miles per week (Answers may vary somewhat depending upon how

decimals were rounded.)

PAGE 87
1. Tami
2. 16
3. 14
4. Matt

PAGE 88
1. 9 hrs, 55 min
2. 5:30 PM
3. 1:20 PM
4. 3:15 AM, Saturday
5. 3:30 AM, Thursday

PAGE 89
1. 10 mph
2. 1,160 miles
3. 12 hours
4. 935 miles
5. 60 mph
6. 10:00 PM

PAGE 90
1. $\frac{16}{20} = \frac{x}{80}$; $x = 64$
2. $\frac{2}{7} = \frac{28}{x}$;

$x = 98$
3. $\frac{900}{2} = \frac{x}{12}$; $x = \$5400$
4. $\frac{7}{9} = \frac{x}{459}$; $x = 357$
5. $\frac{350}{500} = \frac{x}{10}$; $x = 7$
6. $\frac{11}{15} = \frac{x}{180}$ $x = 132$